Favouring a Demonised Plant

Khat and Ethioian smallholder enterprise

Gessesse Dessie

NORDISKA AFRIKAINSTITUTET, UPPSALA 2013

INDEXING TERMS:
Ethiopia
Plant production
Drugs of abuse
Khat
Commercial farming
Small farms
Smallholders
Income
Livelihood

ISSN 0280-2171
ISBN 978-91-7106-731-9
© The author and Nordiska Afrikainstitutet 2013
Production: Byrå4
Print on demand, Lightning Source UK Ltd.

Contents

List of figures and tables

Acknowledgement

This publication is part of an International Foundation for Science (IFS) funded project (S/3599-2) and a NAI fellowship facilitated the writeup of this paper.

Foreword

This publication investigates an interesting agricultural product, khat, an evergreen tree cultivated in parts of Ethiopia for of its fresh leaves, which are chewed for their euphoric properties. The study identifies khat in agricultural landscapes, but more importantly addresses the spatial flow of the khat trade and the agricultural value chains connected with the crop from producer to final consumer, the latter often located in Europe and distant countries. The dynamics of the value chain are analysed in terms of employment, income generation and financial flows and of smallholder-led improvements to khat production in different agricultural landscapes. Such improvements include technical change, innovations and adaptations, capital investments and institutions. The article is testimony to the dynamics of smallholder production and a specific value chain.

Ethiopia is not only a country dominated by agriculture, it is a country of small-holder farmers. The livelihood decisions of this predominant farming community are largely constrained by diminishing land availability, declining soil productivity, the marginalisation of time-tested crops, poor access to technology and the volatility of agricultural markets (Fenta and Ali 2003). In a society so reliant for survival on agricultural land with few other livelihood options, these constraints are the more pronounced. The smaller the amount of available land, the more complicated it becomes for farmers to maintain their customary diversified cropping regime. Under such conditions, farmers are forced to prioritise crops and intensify management to optimise benefits (Rahmato 2009). Often, crops with a high cash return take centre stage and farmers become reliant on the cash to access food. Khat (*Catha edulis* Forsk) is just such a crop.

Khat is an evergreen tree cultivated for the production of fresh leaves that are chewed for their euphoric properties. This plant of eastern Africa and the Arabian Peninsula is a controversial crop, for its consumption is considered a form of drug abuse, while it is at the same time a strongly preferred smallholder crop. Over the past century, khat has emerged from being an obscure crop with limited commercial value to an export earning hundreds of millions of dollars. Khat exports from Ethiopia are expanding rapidly: during 2009-2010 alone, the export value of this plant increased by 51 per cent, while coffee increased 40 per cent and leather products were down by 25.4 per cent (*FORTUNE* 2010). With its emergence on the global scene, a discourse on prohibiting its cultivation and use grew up, and some European, North American and Arabian countries have outlawed it (Andersson *et al.* 2007). However, Ethiopian law on the issue of khat is in limbo, neither supporting nor denouncing its use. Media, community leaders and law enforcement establishments condemn the impact of khat abuse on social well-being. From the farmers' perspective, their choice of the plant is largely attributed to ever-increasing demand for it as a result of growing markets elsewhere and various production constraints at local levels (Gebissa 2010). Historical events, global drug narratives, sub-regional conflicts and food security issues have intensified the khat debate worldwide (Klein 2008; Carrier 2009).

Existing scholarship, emphasising health, sociology and, more recently, political economy, has probably overlooked the most salient reason for the rapid expansion of this crop, namely the contribution it makes to the livelihoods of the producers. This paper explores how important the contribution of this crop is to these livelihoods.

Unlike other agricultural crops, khat is unique in having four characteristics: it is seen as a harmful drug; its chemical constituents are volatile and the plant is easily perishable; it is a valuable commodity; and it is associated with smallholder farming ventures. Khat is highly perishable, its potency degenerating within 48 hours of the cutting of the twigs and leaves (Cassanelli 1986). Its stimulation effects are similar to those of amphetamines and it has been classified by the World Health Organisation as a drug whose abuse can produce mild to moderate psychological dependence (UNODCCP 1999). Smallholders are changing their farming strategies in its favour, and its cultivation influences landscape dynamics, growers' livelihoods and food security (Feyisa and Aune 2003).

Livelihood studies become more complete with the adoption of a sustainable livelihood approach. Resilience and asset-building over time without undermining the natural resource base constitute a sustainable livelihood (Scoones 1998). The sustainable livelihood framework identifies farmers' capacity to mobilise and develop human, natural, financial, social and physical capital. Furthermore, the framework addresses farmers' pursuit of increased income and wellbeing, reduced vulnerability, improved food security and sustainable use of natural resources. To understand khat from a sustainable livelihood perspective, four issues will be addressed: 1) khat geography, including the types of farmers engaged and their different crop management regimes, 2) khat income, embracing the various forms of income the crop generates at different levels and scales; 3) khat employment, specifically the diversity and scope of employment khat production can create; and 4) khat and rural transformation, that is, the khat production process, innovation, knowledge and institutions.

Multiple methods are employed to capture the relevant data, including systematic countrywide visits to khat-growing areas, price and market surveys, interviews with individual farmers and discussions with groups of farmers and use of secondary information. Comparative studies and relative value analysis were employed to summarise the information collected.

Khat is a plant native to Ethiopia that has been consumed for several centuries for mental and physical stimulation. Its commercialisation started at the beginning of the 20th century in eastern Ethiopia, and other growing regions soon followed suit. Today, according to the Central Statistical Agency (CSA) (2010), over 160,000 hectares of land are covered by khat farms and over two million farmers produce khat in all regional states of the country. This crop grows in a wide range of agro-ecological zones between 1,500 and 2,700 metres above sea level. It is mainly cultivated by smallholder farmers on an average of less than one-tenth of a hectare. Figure 1 shows the distribution of major commercial khat production in Ethiopia. The numbers associated with growing regions indicate the period during which khat emerged as a commercial crop, with 1 being the earliest and 7 the most recent. Khat is sold in almost all concentrated settlement areas in Ethiopia, but the amount of khat collected and traded depends on the proximity of farming areas. There are three export centres in the country, Dire Dawa, Jijiga and Addis Ababa, which send khat to Djibouti, Somalia, the UK and China.

Spatial flow of khat trade

From farm gate in rural areas to consumers in urban areas, the khat trade follows a similar pattern. In general, there are five nodes in the spatial flow of the

Figure 1. Commercially significant khat-growing regions in Ethiopia. The map is produced by superimposing the growing regions identified during field visits on existing roadmaps of the country.

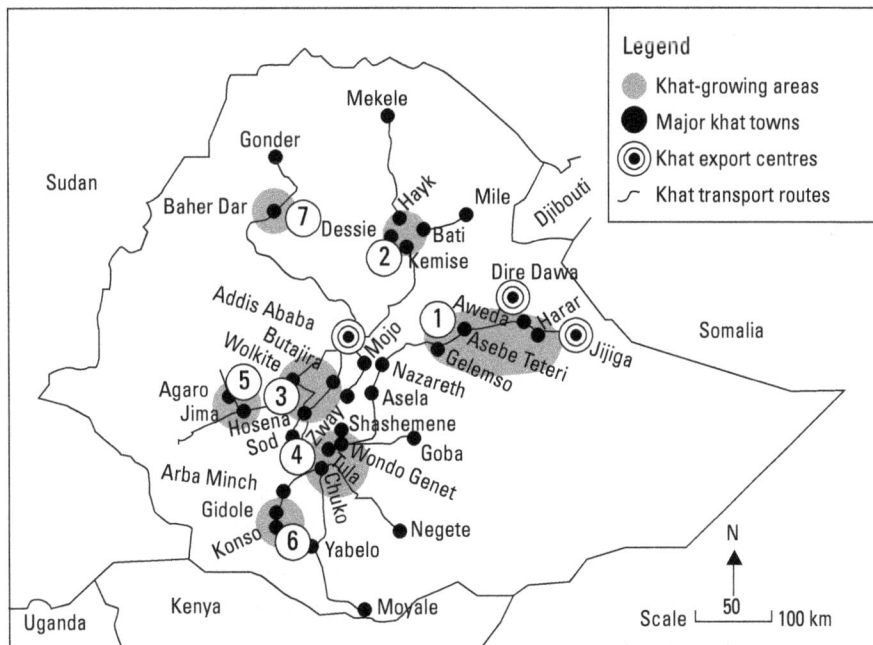

Figure 2. Khat product flow and khat market linkages from the farmers' fields to retailers.

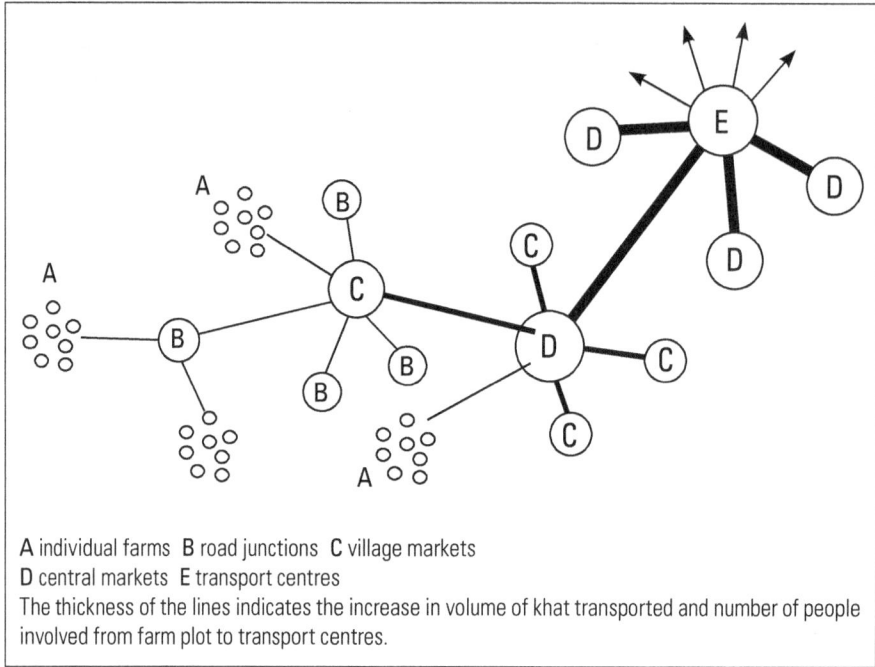

A individual farms B road junctions C village markets
D central markets E transport centres
The thickness of the lines indicates the increase in volume of khat transported and number of people involved from farm plot to transport centres.

khat trade, namely farms, road junctions, village markets (collection hubs), central markets and transport centres (see Figure 2). At each node, two types of sale occur: retail for local consumers located close to trading places and wholesale, to be transported to consumers elsewhere. Spatial flows of khat traded for export are slightly different, in that the companies that collect from villages carry the product all the way to the export centres.

At each node, the volume of khat increases, and as the distance from farms increases, the means of transport change and number of people employed grows. From farms to road junctions, the important means of transport are humans, horse carts or tri-wheelers. At trading junctures, khat is collected, trimmed, watered and packed. At each node, small traders sell goods and services such as food, soft drinks, tea/coffee, cigarettes and plastic bags. In addition, residual materials in the form of leaves and soft branches supply animal feed along the way.

Khat in agricultural landscapes

Today, khat production is part of the wide agro-silvi-pasture complex of Ethiopian rural landscapes. At farm level, khat shares space with food and tree crops and contributes cash to the household economy. As summarised in Table 1, khat is already an important crop in at least seven main growing areas. Its coexistence with enset and eucalyptus in Sidama (growing region 4) and Guraghe

Table 1. Components of agricultural landscapes in khat-growing regions of Ethiopia and their relative importance within farmer homesteads.

Crops	Major khat-growing regions in Ethiopia						
	1	2	3	4	5	6	7
Khat	XX	XX	XX	XX	XX	XX	XX
Coffee	XX	X	X	XX	XX	X	X
Eucalyptus	X	X	XX	XX	X	X	XX
Fruit	XX		X	XX	X		
Enset			XX	XX	X		
Cereals	X	X	X	X	XX	X	XX

Importance of crops designated by X for subsistence and XX for cash.
Source: Interviews and discussions with khat farmers in khat-growing regions.

(growing region 3) indicate unique food, wood and cash integration. Enset is a banana-like food crop strongly linked to pastoralism, and its resistance to moisture deficiencies means it supports a high population density in southern Ethiopia. Eucalyptus is a fast-growing tree, able to coppice after harvest, and is an important source of wood for construction in the country. Elsewhere, khat combines well with food crops such as maize, sorghum and teff in Baher Dar and Harar (growing regions 7 and 1 respectively). These areas also produce eucalyptus. Coffee, one of the country's most important export crops, co-exists with khat in differing degrees in all growing regions .

In addition to the mainly monocropping type of khat cultivation, two types of mixing are widely practised, line intercropping and patch-cropping. Intercropping is practised in growing regions 1 and 5. In Harar (growing region 1), mainly sorghum or maize is intercropped in the narrow space of about a metre between khat bushes. By contrast, in Jimma (growing region 5), farmers maintain wide spacing of up to 4 metres between khat plants and intercrop teff. Patch-cropping, whereby pieces of a farmers' landholding are allocated to different crops, is the most dominant form of khat management in Ethiopia.

Financial benefit

Export wise, khat represents the fourth most important commodity in the 10 years up to 2007 (Figure 2). Thereafter, the export of khat overtook hides and skins, one of Ethiopia's major export items from its cattle resources, the most numerous in Africa. In 2010 alone, khat fetched over US$200 million and khat tax revenue could have reached over half a billion birr (US$289 million).

Income from khat is important at regional, zonal and district levels. According to the finance bureau in Guraghe Zone, between 1999 and 2009 the average annual tax revenue was over 13.5 million birr. The maximum and minimum revenue per year during that time was 21.6 and 7.9 million birr respectively. In Wondo Genet, the district finance bureau reported that between 2007 and

Figure 3. Value of Ethiopian export commodities as a percentage of coffee per annum

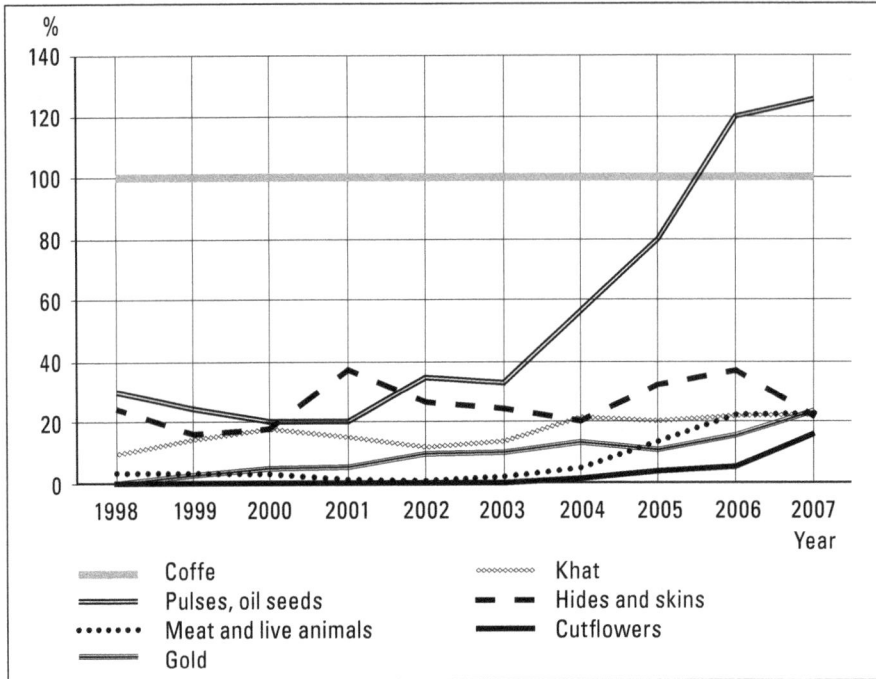

Source: Calculated from records of the Ethiopian revenue and customs authority

2009 the average annual tax income has 7.3 million birr, with a maximum of 9.3 million and minimum of 6.0 million per annum.

At the farm level, income per hectare from khat surpasses all major agricultural crops by several margins: it is 14.5 times more than for grain/cereals; 17 for pulses; six for oilseeds; and four for coffee (Table 2). In all other measures, including total production area, number of producing farmers, annual production, land productivity and landholding per farmer, khat represents a small fraction of the other crops.

Table 2. Relative value of khat with regard to area covered, growers involved, production, income and landholding in Ethiopia

Measures	% khat to Grain/cereals	% khat to Pulses	% khat to Oilseeds	Khat (absolute figure)	% khat to Coffee
Area (ha)	0.8	10.8	23.1	163,227	40.1
Producers (farmers)	9.2	32.9	74.2	2,232,397	63.8
Production (quintal)	0.5	7.7	22.2	1,368,027	50.0
Productivity (quintal/ha)	55.3	71.3	96.1	8	124.8
Price (birr/quintal)	2,623.3	2,405.3	625.0	15,000	333.3
Income (birr/ha)	1,450.7	1,716.2	600.3	125,716	416.0
Landholding (ha)	7.9	32.7	31.1	0.07	62.8

Figures under each crop in each cell indicate the value of khat relative to the crop in that column (US$1 = 17.2 birr).
Source: Calculated and summarised from price list at Addis Ababa Market for Ethiopian Grain Trade Enterprise (EGTE 2009) values and agricultural sample survey 2008–2009 results (CSA 2009)

Khat-related employment

There is a multitude of actors involved in the khat production process, and the process itself is diverse, including direct, indirect and derivative activities. Direct activities include khat transactions, khat sorting and packing and khat transporting. Indirect activities include several forms of small trading ventures that benefit from khat production and marketing. Derivative activities are production processes set in motion by the existence of khat, such as production of groundnuts, bottled water and cigarettes. Hence, khat-production activities provide employment and cash for several people, not just farmers

Closer analysis of a case from Butajira (growing region 3) reveals that khat production involves at least six major sets of activities, each of which has multiple sub-activities and employs about 17 people in total (Table 3). The six major activities include hauling/feeding, auction/bulk trade, sorting/packing, delivery/transport, retail and residual. Table 3 shows that over 400 individuals of different genders and age groups and from various sources are involved every day

Table 3. Employment in the khat trade and cash gain by different actors in Butajira khat market

		Gender		Age				Source		Income
	Actors	M	F	O	M	Y	C	R	U	birr/p/d
1	Farmers	150			150			150		666.7
2	Auctioneers	20			20				20	250.0
3	Recorders	20				20			20	50.0
4	Rope vendors	3					3		3	16.7
5	Wrapping-leave vendors		3		3				3	50.0
6	Soaking-water vendors	4					4		4	7.5
7	Sorters and wrappers	10	10						10	15.0
8	Bundle binders	5					5		5	10.0
9	From farm carriers	15				15		15		20.0
10	Horse cart drivers	40			40				40	20.0
11	Tri-wheeler drivers	30			30				30	10.0
12	Wooden cart pushers	5					5		5	10.0
13	Boiled coffee vendors		10			10			10	50.0
14	Peanut vendors		15		15				15	20.0
15	Residual collectors		5	5					5	10.0
16	Garbage cleaners	7				7			7	21.4
17	Minibus drivers	10			10				10	100.0
18	Bulk buyers	50	5		30	25		10	45	
	Total	369	38	15	228	147	17	175	232	

There are four age categories:
O old >50 years
M mid >30 to 50 years
Y young teenagers
C children under 10 years.
Source refers to rural (**R**) and urban (**U**). Income is in birr per person per day.
Source: Survey in Butajira khat auction market

in running the khat trade. In this male-dominated activity, the proportion of women is less than 10 per cent. Age-wise, middle-aged men are the most numerous, followed by youngsters. Elders and children have the lowest involvement. As to the urban-rural divide, over 57 per cent of participants are urbanites.

The average daily income of actors ranges between 7.5 birr to 250 birr. The minimum income is 2.5 birr short of the lowest government salary for an eight-hour working day, while the maximum daily income is more than twice the salary of a university professor. The income each actor gets is for about three hours of activity between 10 am and 1 pm, during which time the khat auction is in swing.

Employment associated with khat is either spatially permanent or semi-permanent. The Butajira khat auction market shows that, apart from the farmers

Figure 4. Spatial setting of khat trade network at Butajira khat auction market near the bus terminus
Source: Mapped from survey of the site during 2010

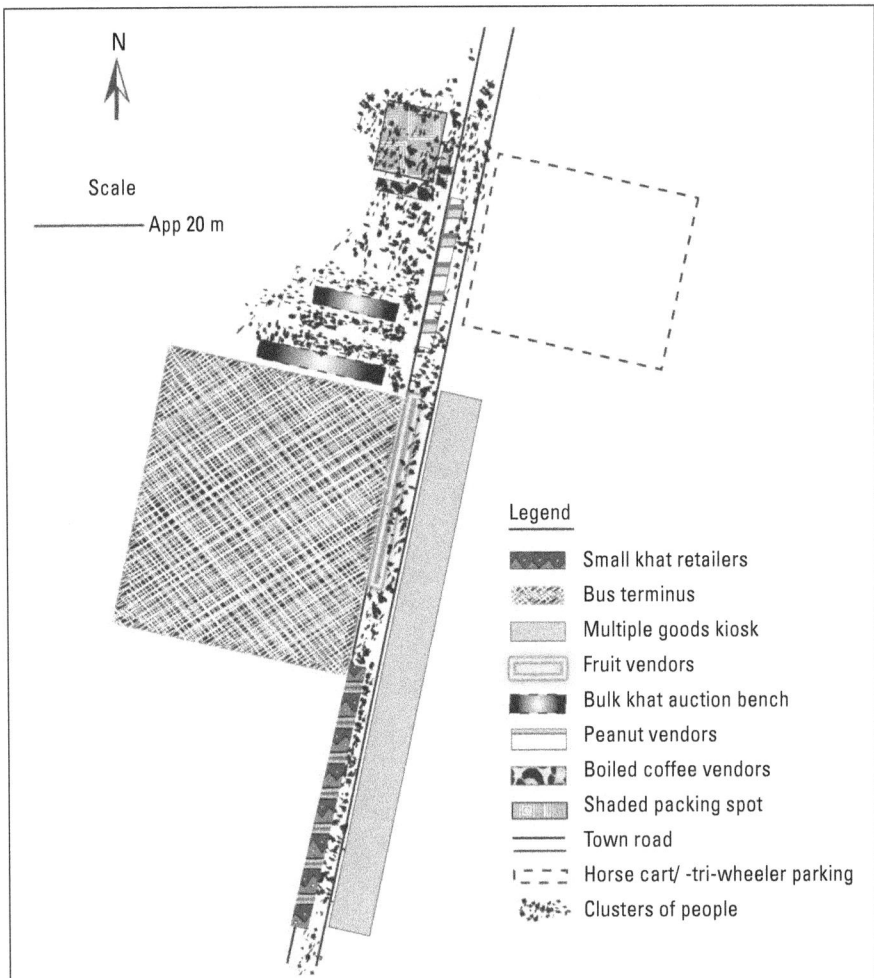

N

Scale

App 20 m

Legend

Small khat retailers
Bus terminus
Multiple goods kiosk
Fruit vendors
Bulk khat auction bench
Peanut vendors
Boiled coffee vendors
Shaded packing spot
Town road
Horse cart/ -tri-wheeler parking
Clusters of people

who come two or three times a year following the harvesting of their leaves, the remaining actors involved in the trade are regulars. Their space in the market has a certain permanence and is subject to fees and rent based on location. The auction benches are subdivided and owned by different individuals with the skills to lead the trade and to win the trust of both farmers and traders. Vendors occupy the same place every day, as long as they pay certain fees. For example, each coffee seller pays two birr per day to the person who owns the shaded spot. In contrast to this informal occupancy of space, multipurpose goods sellers own their kiosks or rent them through a formal contractual agreement.

Smallholder farmers confront several production constraints, particularly khat farmers, who get no government support of any kind. Interviews with farmers reveal there are at least nine production constraints: moisture, wetlands, slopes, disease, cultivation, management, harvesting, theft and markets. Farmers seek to overcome these constraints or at least make the best possible efforts to reduce their impact. Table 4 is based on a countrywide survey of khat farmers' endeavours to overcome production constraints by adopting agricultural technologies and fashioning innovating mechanisms, creating landesque capital and developing various institutions.

Except in growing region 5 and one locality in growing region 4, where landholding per household is comparatively large at about a hectare, farmers from all growing regions report land as being among the major limitations to cultivating khat. As a result, farmers tend to move to formerly unproductive lands such as steep slopes, natural forest and wetlands. Farmers in growing region 1 and 2 control slopes with well-constructed permanent stone bands and terraces (see Figure 4). In Wondo Genet, growing region 4, farmers drain wetlands by planting eucalyptus and digging drainage ditches to make the soil suitable for cultivating khat.

Khat farmers select and apply traditional knowledge and practices to protect their valuable crop. In growing regions 2 and 7, water pipes cut locally from plastic or canvas are commonly used to pump water for irrigation over hundreds of metres. With regard to theft, in addition to community policing, farmers benefit from age-old sorcerers' spirit warnings. Additionally, they use live hedges of thorny plants to keep humans away and poisonous plants to keep animals out.

Sometimes, farmers adopt unique measures to overcome serious constraints. A farmer in Hayk, growing region 2, reported that he attracts termites under his khat bushes by feeding them hay so as to moisten his plants with their exudates. Khat plants benefiting from termites in this way show a healthy green colour, unlike bushes deprived of moisture. However, while the termites are good for the plants, they are not for the foundations of the farmer's house.

Another interesting innovation in this area involves the control by some farmers of insect-breeding on their khat plants in order to improve the quality of the leaves by altering their moisture content. High moisture in leaves substantially reduces the quality and hence the price of khat. Elsewhere, such insects are seen as undesirable by farmers and killed with chemical insecticides to enhance the growth of succulent leaves and twigs. The application of chemical insecticides is common practice in all khat-growing areas of the country.

With regard to cultivation, farmers near Jima, growing region 5, modified the size of the ox-yokes used for ploughing. To allow for in-line intercropping of khat and food crops and for greater ease of movement, the farmers found it necessary to reduce the size of the yoke. This modification is interesting in that it has been made to a farming implement that is several millennia old.

The fact that khat is perishable motivates farmers to employ several methods to ensure that the trade is efficient and speedy. In all khat regions, moist wrapping is used to keep the khat fresh, the transport used is fast so that markets can soon be reached and the trade is tied to social capital and customary laws. Mobile phones are important communication tools in all growing areas and are used to enhance the local and regional khat trade.

Table 4. Farmer-led improvements of the khat-production process

Production constraints	Technology	Innovations/Adoption	Landesque capital	Institution
Moisture	• Water harvesting • Pump irrigation • Channel irrigation	• Termite culture • Improvised pipes • Channel pipe combined	• Ponds • Irrigation canal	• Water sharing/ distribution
Wetland	• Water draining • Field drying	• Planting aggressive trees • Ditch-draining	• Ditches	• Common management
Slope	• Levelling • Soiling	• Wide-bench terrace • Gully filling	• Stone bands/ terraces	• Waterway/path common use
Diseases	• Pesticide • Traditional methods	• Improvised sprayer • Smoking, inoculation		
Cultivation	• Intercropping • Fertilisation	• Spacing, tool modification • Mulch, manure		
Management	• Tree management • Shrub management	• Climb harvesting • Frequent harvesting		
Harvesting	• Branch bundle • Twig bundle	• Moist wrapping • Fast transaction		
Theft	• Local spirits • Fencing • Nearby residence	• Spirit fencing/ scarecrows • Thorny, poisonous fence	• Settlement areas	• Community police
Market	• Mobile telephone • Speedy cars	• Night market • Auction markets • Telemarketing	• Footpath, motor road	• Social network • Customary law

Source: Records of field visits to khat-growing regions and markets in Ethiopia

Figure 5 shows a stone band designed and constructed by farmers in northern Ethiopia specifically for khat farming. Unlike other stone bands in this part of Ethiopia, which are financially and technically supported by government extension services, this is a local farmer enterprise. Here, the bench has been widened to enable oxen to turn during ploughing, thereby overcoming one of the limitations common to such structures. These stone bands are common in growing regions 2 and 1.

Figure 5. Stone bands constructed by farmers in Kemise, growing region 2, for khat production. Observe the khat bushes and the farmer ploughing the land with oxen

The results show that khat contributes to multiple livelihood opportunities for the actors involved in its production. This result is consistent with the findings of Carrier (2007) that the livelihood contribution of khat is not only for those directly involved in its trade. Unlike food crops produced and consumed locally, khat is exported in return for foreign currency and tax revenue. The fact that it is a smallholder venture and is expanding through all farming systems indicates its importance to cultivators and their land use.

The emergence of khat may intensify land use conflicts, but does not necessarily mean the plant takes over land on which other crops were produced. The tendency to intercrop khat and other crops indicates the farmers' strategy to simultaneously produce for food and for cash (Feyisa and Aune 2003). With landholdings of below half a hectare per household, farmers may not be able to produce sufficient food without improved land productivity, agricultural technology and credit. Khat seems to enable them overcome such constraints by producing a high income per small land area (Gessesse and Kinlund 2008). Farmers with larger landholdings seem to maintain their food crops and at the same time allocate sufficient land for khat production.

The money actors make indubitably prompts more khat production and tempts farmers to prioritise khat over food crops. Additionally, the sense of complacency the money may bring about, particularly with regard to access to food, may lead to the marginalisation of food crops. On the other hand, scepticism regarding khat and seasonal price volatility seems to restrain farmers from cultivating khat on all their land. Unless food crop production is futile despite the investments farmers make in it and food is available nearby, it is unlikely that farmers will stop planting food crops on their land for good. Such behaviour runs counter to the general risk-averse strategy adopted by rational farmers (Milich and Al-Sabbry 1995).

Khat production processes seem to touch upon all livelihood assets, namely human, social, physical, natural and financial capital. Farmers exhibit dependable conduct with regard to cultivation, management and protection. The market is run by specific role-playing actors through networks and the cohesion of the networks remains strong due to the mutual benefit they confer. Owing to the perishability of the plant, speed is crucial and cannot be achieved without operating logistics, efficient trade and social capital. The physical capital is mainly in the form of landesque capital created to intensify cultivation. Most khat-growing regions display characteristics similar to indigenous systems, according to Widgren and Sutton (2004), and achieve intensive exploitation of all usable land by using terrace walls to conserve soil and hill-furrow irrigation to supplement low or seasonal rainfall. The formation of such capital not only improves the productivity and value of the land, it also indicates the sense of security and confidence farmers have about their khat lands. However, the impact of khat production on maintaining natural capital is debatable, given that khat farms certainly affect soil productivity, biodiversity and natural vegetation (Krikorian 1985; Msuya and Madoffe 1999).

Monocrop management can have negative implications for the agro-biodiversity of the agricultural landscapes of the khat-growing regions. With regard to soil, both negative and positive implications are reported. The permanency of khat bushes and accompanying soil conservation structures positively impact the soil, but the constant and frequent mining of soil nutrients by the leaves harvested can be negative.

Financial capital seems the most obvious return on khat production, particularly because khat is superior in this respect to all annual and perennial agricultural crops and trees (Table 2). It is not only farmers who benefit by deriving a cash income, but rural unemployed youths also benefit from khat ventures. With financial transactions centred on khat production nodes, namely farm gate, village markets and urban areas, opportunities are created for petty trade. The inputs needed in khat production processes give rise to business opportunities for women, youths, children and entrepreneurs.

Khat production requires casual labourers, transporters and traders. It entails multiple processes at different scales and levels, thereby creating an important trickle-down effect. Despite its economic importance, the government's role is unclear, with the state neither officially supporting khat production nor providing the agricultural packages available for other crops. The Ethiopian government, while it honours khat traders for their significant role in securing foreign currency, thereby implying state interest, provides no support for exports or improved yields. The state's tax-collecting mechanisms include the ministry of finance, municipalities and tax and revenue authorities. While religious groups, community leaders, law enforcement officials and politicians remain apprehensive about khat, the number of consumers is increasing all the time.

It seems that land size and intended management type affect khat cropping. However, interesting intercropping approaches are apparent in some growing areas of the country where khat is mainly a monocrop venture. Khat is different from coffee, for which shade trees are necessary, and sugarcane, which is strictly a single crop, in that it is usually selectively combined with other crops. The fact that it is a non-food plant managed for income can affect food production at household level. The high income khat provides can help farmers' purchase food, but price fluctuations and lack of own food is likely to create insecurity.

One may argue that as long as income from khat is sustainable, food can be bought, and that furthermore such an income can motivate farmers to save and invest. There are at least two counterarguments: food production elsewhere may not be sufficient even if farmers can afford to buy and khat prices are volatile, resulting in income declines and an inability by farmers to save and invest. As in most cash crop-reliant regions of the country, farmers tend to turn to consumerism and become complacent about the relative reliability of their income season after season. It seems that the livelihood improvements of the farmers derive not only from the good cash income khat generates, but also from the ability to maintain the income in bad days and food deficit seasons. Certainly, khat-driven technologies, innovations and institutions strengthen livelihoods by improving human, physical, social, natural and financial capital. Of course, the sustainability of the khat-driven benefits is brought into question by the controversial nature of the crop and its possible banning.

Income from khat, unlike other crops, has several aspects. It is to be acknowledged that while this is beneficial to actors involved in khat production, the income of consumers is drained. The income reaching a household depends on the size of the khat farm, the season of the harvest, the marketability of the crop and the state of the infrastructure, such as roads and communications. However, even if these conditions are met, farmers do not necessarily derive full market benefit, because intermediaries are also involved. It is true that there is trust and mutual benefit in the trade systems, but judging by the increase in price as distance increases, traders enjoy a profit margin much higher than the modest value added by their transporting the crop.

Some studies assert that growth in cereals and other staple crops should be given priority in decreasing poverty in Ethiopia (Diao and Pratt 2007). The expansion of khat production runs counter to this assertion and increased production of staple crops seems to be less of a priority among smallholder farmers. Even at national and regional levels, income from khat appears more important, as it may eventually improve farmers' livelihoods by enhancing development in general. On the other hand, khat also spawns corruption and monopolies. Even so, some of what profit remains flows into government coffers as tax and can be used by government to supplement local growth. There is thus a trickle-down, however limited and often negligible, of national and regional revenues into the livelihoods of smallholder farmers, which improves infrastructure, education, communications and bureaucratic efficiency, all of which are important to boosting the khat trade and production process. This trade, in turn, creates access to markets and also improves food and supply flows.

The flows of plant materials and cash, the networks and employment associated with the khat trade can also have livelihood implications for farmers. Along the spatial gradient, khat leaves a trail of money that constitutes the income of people involved in the business. Material flows in particular, apart from the piles of garbage municipalities complain of, provide otherwise unavailable animal fodder for urban herders. Townspeople who have limited livelihood options, or those who want to diversify their income sources, keep goats, sheep and cows, and khat residuals, leaves and wrappings, help them in this regard.

As with most agricultural enterprises, khat creates employment. Employment is an aspect of khat-generated income. What is different is that khat creates additional opportunities for unemployed rural youths and poor farmers whose livelihood options are limited: it is not only the khat farmers and traders who gain employment. Because of the perishability and high value of the commodity, the trade requires several actors and functional logistics, as well as other service providers. To achieve the necessary pace, not only fast transport matters. A logistic network involving diverse actors playing different roles needs to be in place. Middlemen, however different in character, including auctioneers, are needed to link the farms to the buyers, benefiting both farmers and buyers and maintaining a smooth flow of the products involved in the trade.

The high volume of cash flow at different points in the khat network attracts people who seek to benefit from it. Entering this business and gaining paid employment in it requires no specific knowledge. In most cases, trust and association are sufficient, and power is seldom necessary. The khat trade has several components that anyone can become involved in, including children, women and people of different age groups. Children are involved in less physically demanding activities, and are apparently accepted into the business out of sympathy for their plight as street children and beggars. Coffee vendors are important in countering insomnia, while groundnut traders' benefit from the increasing trend of chewing nuts combined with khat.

Khat production differs from that of other crops, both cash and subsistence. Khat, which is consumed fresh, requires prompt handling and working logistics. Other agricultural crops produced by smallholders do not need this: coffee, for instance, requires picking, drying and threshing, all of which can be done by the farmer, while maize needs only harvesting and selling.

It is important to note that income derived from khat isn't always positively viewed even by people making good money from the trade, for the following reasons:

- Children in khat-growing areas don't like to go school when they can make easy money;
- Farmers tend to become consumers, instead of producing their own food and saving;
- Risks of price volatility exist, which are likely to create food deficits in time;
- Continuous income from a perennial crop and frequent harvests of leaves make farmers mainly dependent on khat;
- Powerful, well-connected people are attracted into this lucrative and relatively easy money-making venture, and will eventually monopolise the trade;
- Corruption and farmer marginalisation are common; and
- The uncertain legal status of khat may mean the trade is not sustainably beneficial.

Conclusion

Khat confers significant benefits at national, regional and household levels. The khat production process employs people from diverse sections of society and across gender, age, class as well as urban and rural divides. The income trickles down the rural and urban gradient, affecting people from both sectors. Khat management varies by growing region, and the plant coexists with other crops. Farmers adopt different technologies, innovate, develop landesque capital and various institutions to cultivate khat in the face of various constraints.

Khat, dubbed a social ill for many, is at the same time a source of livelihood for many others. The negative health, social and cultural impacts on consumers should not be overemphasised, as it is possible to minimise the impact by controlling the rate of consumption and abuse, as with other accepted drugs such as alcohol and tobacco. On the other hand, khat should not be viewed as the only dependable source of livelihood. Farmers need to consider additional livelihood options, diversify, save and invest. However, the producers' preference for khat, despite the perishability controversies surrounding the crop, shows just how important khat is to farmers.

Market places are not sites legally designated by municipalities, but are constantly selected by the actors involved. Chosen areas are bus terminals, taxi terminals, easily accessible town quarters and so on. The unclear use rights to the spaces makes khat farmers vulnerable to any land use changes municipalities may impose on khat markets. Government involvement here is just to collect tax, with no improvements in services and goods.

References

Anderson, D., S. Beckerleg, D. Hailu and A. Klein, 2007, *The Khat Controversy: Stimulating the Debate on Drugs*. Oxford: Berg.

Carrier, N., 2009, "Khat in the Western Indian Ocean: Regional Linkages and Disjunctures", *Etudes Océan Indien* 42-43, 2009.

Carrier, N.C.M., 2007, *Kenyan Khat the social life of a stimulant*. Leiden: Brill.

Cassanelli, L.V., 1986, "Qat: Changes in the production and consumption of a quasi legal commodity in northeast Africa", in Arjun Appadurai (ed.), *The social life of things, commodities in cultural perspective*. Cambridge: Cambridge University Press.

CSA (Central Statistical Agency), 2010, *Agricultural Sample Survey 2008/2009. Report on Area and Production of Crops*. Statistical Bulletin 446, Volume 1, Addis Ababa.

Diao, X. and A.N. Pratt, 2007, "Growth options and poverty reduction in Ethiopia an economy-wide model analysis", *Food Policy* 32:205-28.

Fenta, T. and O. Ali, 2003, Proceedings of the Food Security Conference 2003, Challenges and Prospects of Food Security in Ethiopia, 13-15 August, Addis Ababa.

Feyisa, T.H. and J.B. Aune, 2003, "Khat expansion in the Ethiopian highlands. Effects on the farming system in Habro District", *Mountain Research and Development* 23(2): 185-9.

FORTUNE, 2010, A weekly business newspaper in Ethiopia. Vol. 11, No. 535, Sunday, 1 August.

Gebissa, E., 2010, *Taking the Place of Food, Khat in Ethiopia*. Trenton NJ: Red Sea Press.

Gessesse D. and P. Kinlund, 2008, "Khat expansion and forest decline in Wondo Genet, Ethiopia", *Geografiska annaler*, Series B, Human Geography 90 (2):187-203.

Klein, A., 2008, "Khat in the Neighbourhood – Local Government Responses to Khat Use in a London Community", *Substance Use and Misuse* 43:819-31.

Krikorian, A.D., 1985, "Growth mode and leaf arrangement in Catha edulis (kat)", *Economic Botany* 39 (4):514-21.

Milich, L. and M. Al-Sabbry, 1995, "The Rational Peasant vs. Sustainable livelihoods: The case of Qat in Yemen", http://www.togdheer.com/khat/rational.shtml. Accessed 15 November 2008.

Msuya, T.S. and S.S. Madoffee, 1999, "The effect of Catha edulis leaf harvesting: Case of west Usambara mountains, Tanzania", in N.D. Burgess, M. Nummelin, J. Fjeldsa, K.M. Howell, K. Lukumbyzya and L. Mhando (eds), *Biodiversity and conservation of the Eastern arc mountains of Kenya and Tanzania*. Nairobi: National Museums of Kenya.

Rahmato, D., 2009, *The peasant and the state. Studies in agrarian change in Ethiopia 1950s-2000s*. Addis Ababa: Addis Ababa University Press.

Scoones, I., 1998, "Sustainable rural livelihoods: A framework for analysis", Working Paper 72, IDS, Brighton.

UNODCCP, 1999, "The drug nexus in Africa, ODCP studies on Drugs and Crime", Monograph No. 1, United Nations Office for Drug Control and Crime Prevention, Vienna, http://www.unode.org/pdf/report_1999_03_01_1.pdf. Accessed 15 November 2008.

Widgren, M. and E.G. Sutton, 2004, *Islands of Intensive Agriculture in Eastern Africa.* Oxford: James Currey.

CURRENT AFRICAN ISSUES PUBLISHED BY THE INSTITUTE

Recent issues in the series are available electronically
for download free of charge www.nai.uu.se

1. *South Africa, the West and the Frontline States. Report from a Seminar.* 1981, 34 pp, (out-of print)

2. Maja Naur, *Social and Organisational Change in Libya.* 1982, 33 pp, (out-of print)

3. *Peasants and Agricultural Production in Africa. A Nordic Research Seminar. Follow-up Reports and Discussions.* 1981, 34 pp, (out-of print)

4. Ray Bush & S. Kibble, *Destabilisation in Southern Africa, an Overview.* 1985, 48 pp, (out-of print)

5. Bertil Egerö, *Mozambique and the Southern African Struggle for Liberation.* 1985, 29 pp, (out-of print)

6. Carol B.Thompson, *Regional Economic Polic under Crisis Condition. Southern African Development.* 1986, 34 pp, (out-of print)

7. Inge Tvedten, *The War in Angola, Internal Conditions for Peace and Recovery.* 1989, 14 pp, (out-of print)

8. Patrick Wilmot, *Nigeria's Southern Africa Policy 1960–1988.* 1989, 15 pp, (out-of print)

9. Jonathan Baker, *Perestroika for Ethiopia: In Search of the End of the Rainbow?* 1990, 21 pp, (out-of print)

10. Horace Campbell, *The Siege of Cuito Cuanavale.* 1990, 35 pp, (out-of print)

11. Maria Bongartz, *The Civil War in Somalia. Its genesis and dynamics.* 1991, 26 pp, (out-of print)

12. Shadrack B.O. Gutto, *Human and People's Rights in Africa. Myths, Realities and Prospects.* 1991, 26 pp, (out-of print)

13. Said Chikhi, Algeria. *From Mass Rebellion to Workers' Protest.* 1991, 23 pp, (out-of print)

14. Bertil Odén, *Namibia's Economic Links to South Africa.* 1991, 43 pp, (out-of print)

15. Cervenka Zdenek, *African National Congress Meets Eastern Europe. A Dialogue on Common Experiences.* 1992, 49 pp, ISBN 91-7106-337-4, (out-of print)

16. Diallo Garba, *Mauritania–The Other Apartheid?* 1993, 75 pp, ISBN 91-7106-339-0, (out-of print)

17. Zdenek Cervenka and Colin Legum, *Can National Dialogue Break the Power of Terror in Burundi?* 1994, 30 pp, ISBN 91-7106-353-6, (out-of print)

18. Erik Nordberg and Uno Winblad, *Urban Environmental Health and Hygiene in Sub- Saharan Africa.* 1994, 26 pp, ISBN 91-7106-364-1, (out-of print)

19. Chris Dunton and Mai Palmberg, *Human Rights and Homosexuality in Southern Africa.* 1996, 48 pp, ISBN 91-7106-402-8, (out-of print)

20. Georges Nzongola-Ntalaja *From Zaire to the Democratic Republic of the Congo.* 1998, 18 pp, ISBN 91-7106-424-9, (out-of print)

21. Filip Reyntjens, *Talking or Fighting? Political Evolution in Rwanda and Burundi, 1998–1999.* 1999, 27 pp, ISBN 91-7106-454-0, SEK 80.-

22. Herbert Weiss, *War and Peace in the Democratic Republic of the Congo.* 1999, 28 pp, ISBN 91-7106-458-3, SEK 80,-

23. Filip Reyntjens, *Small States in an Unstable Region – Rwanda and Burundi, 1999–2000,* 2000, 24 pp, ISBN 91-7106-463-X, (out-of print)

24. Filip Reyntjens, *Again at the Crossroads: Rwanda and Burundi, 2000–2001.* 2001, 25 pp, ISBN 91-7106-483-4, (out-of print)

25. Henning Melber, *The New African Initiative and the African Union. A Preliminary Assessment and Documentation.* 2001, 36 pp, ISBN 91-7106-486-9, (out-of print)

26. Dahilon Yassin Mohamoda, *Nile Basin Cooperation. A Review of the Literature.* 2003, 39 pp, ISBN 91-7106-512-1, SEK 90,-

27. Henning Melber (ed.), *Media, Public Discourse and Political Contestation in Zimbabwe.* 2004, 39 pp, ISBN 91-7106-534-2, SEK 90,-

28. Georges Nzongola-Ntalaja, *From Zaire to the Democratic Republic of the Congo.* Second and Revised Edition. 2004, 23 pp, ISBN-91-7106-538-5, (out-of print)

29. Henning Melber (ed.), *Trade, Development, Cooperation – What Future for Africa?* 2005, 44 pp, ISBN 91-7106-544-X, SEK 90,-

30. Kaniye S.A. Ebeku, *The Succession of Faure Gnassingbe to the Togolese Presidency – An International Law Perspective.* 2005, 32 pp, ISBN 91-7106-554-7, SEK 90,-

31. Jeffrey V. Lazarus, Catrine Christiansen, Lise Rosendal Østergaard, Lisa Ann Richey, *Models for Life – Advancing antiretroviral therapy in sub-Saharan Africa.* 2005, 33 pp, ISBN 91-7106-556-3, SEK 90,-

32. Charles Manga Fombad and Zein Kebonang, *AU, NEPAD and the APRM – Democratisation Efforts Explored.* Edited by Henning Melber. 2006, 56 pp, ISBN 91-7106-569-5, SEK 90,-

33. Pedro Pinto Leite, Claes Olsson, Magnus Schöldtz, Toby Shelley, Pål Wrange, Hans Corell and Karin Scheele, *The Western Sahara Conflict – The Role of Natural Resources in Decolonization.* Edited by Claes Olsson. 2006, 32 pp, ISBN 91-7106-571-7, SEK 90,-

34. Jassey, Katja and Stella Nyanzi, *How to Be a "Proper" Woman in the Times of HIV and AIDS.* 2007, 35 pp, ISBN 91-7106-574-1, SEK 90,-

35. Lee, Margaret, Henning Melber, Sanusha Naidu and Ian Taylor, *China in Africa.* Compiled by Henning Melber. 2007, 47 pp, ISBN 978-91-7106-589-6, SEK 90,-

36. Nathaniel King, *Conflict as Integration. Youth Aspiration to Personhood in the Teleology of Sierra Leone's 'Senseless War'.* 2007, 32 pp, ISBN 978-91-7106-604-6, SEK 90,-

37. Aderanti Adepoju, *Migration in sub-Saharan Africa.* 2008. 70 pp, ISBN 978-91-7106-620-6, SEK 110,-

38. Bo Malmberg, *Demography and the development potential of sub-Saharan Africa.* 2008, 39 pp, 978-91-7106-621-3

39. Johan Holmberg, *Natural resources in sub-Saharan Africa: Assets and vulnerabilities.* 2008, 52 pp, 978-91-7106-624-4

40. Arne Bigsten and Dick Durevall, *The African economy and its role in the world economy.* 2008, 66 pp, 978-91-7106-625-1

41. Fantu Cheru, *Africa's development in the 21st century: Reshaping the research agenda.* 2008, 47 pp, 978-91-7106-628-2

42. Dan Kuwali, Persuasive Prevention. *Towards a Principle for Implementing Article 4(h) and R2P by the African Union.* 2009. 70 pp. ISBN 978-91-7106-650-3

43. Daniel Volman, *China, India, Russia and the United States. The Scramble for African Oil and the Militarization of the Continent.* 2009. 24 pp. ISBN 978-91-7106-658-9

44. Mats Hårsmar, *Understanding Poverty in Africa? A Navigation through Disputed Concepts, Data and Terrains.* 2010. 54 pp. ISBN 978-91-7106-668-8

45. Sam Maghimbi, Razack B. Lokina and Mathew A. Senga, *The Agrarian Question in Tanzania? A State of the Art Paper.* 2011. 67 pp. ISBN 978-91-7106-684-8

46. William Minter, *African Migration, Global Inequalities, and Human Rights. Connecting the Dots.* 2011. 95 pp. ISBN 978-91-7106-692-3

47. Musa Abutudu and Dauda Garuba, *Natural Resource Governance and Eiti Implementation in Nigeria.* 2011. 74 pp. ISBN 978-91-7106-708-1

48. Ilda Lindell, *Transnational Activism Networks and Gendered Gatekeeping. Negotiating Gender in an African Association of Informal Workers.*
2011. 44 pp. ISBN 978-91-7106-712-8

49. Terje Oestigaard, *Water Scarcity and Food Security along the Nile. Politics population increase and climate change.*
2012. 92 pp. ISBN 978-91-7106-722-7

50. David Ross Olanya, *From Global Land Grabbing for Biofuels to Acquisitions of AfricanWater for Commercial Agriculture.*
2012. 41 pp. ISBN 978-91-7106-729-6

51. Gessesse Dessie, *Favouring a Demonised Plant. Khat and Ethiopian smallholder enterprise.*
2013. 28 pp. ISBN 978-91-7106-731-9

www.ingramcontent.com/pod-product-compliance
Lightning Source LLC
Chambersburg PA
CBHW080210300326
41934CB00039B/3446